Living Australia

FRANK HADDON & TONY OLIVER

Living Australia

Animals and their homes

HODDER AND STOUGHTON
SYDNEY AUCKLAND LONDON TORONTO

First published in 1986 by Hodder & Stoughton (Australia) Pty. Limited, Apollo Place, Lane Cove, NSW 2066, Australia.
© text, Frank Haddon, 1986
© illustrations, Tony Oliver, 1986
National Library Cataloguing-in-Publication entry
Haddon, Frank
  Living Australia
  ISBN 0 340 35960 9
  1. Ecology – Australia – Juvenile literature.
  I. Oliver, Tony, 1940. II. Title.
574.5′0994
Typeset in 14/16 Bembo by G.T. Setters Pty. Limited, Sydney.
Printed by Colorcraft Ltd, Hong Kong.

CONTENTS

Coral reefs

Seashores

Swamps & Estuaries

Eucalypt forests

Rainforests

Open woodlands & Plains

Deserts

Tablelands & Alps

Urban

INTRODUCTION

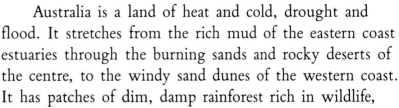

Australia is a land of heat and cold, drought and flood. It stretches from the rich mud of the eastern coast estuaries through the burning sands and rocky deserts of the centre, to the windy sand dunes of the western coast. It has patches of dim, damp rainforest rich in wildlife, with towering figs draped in lianas, and rotting vegetation alive with fungi and microscopic animals. Wide, open plains stretch beyond patches of woodland sheltering bands of lazing kangaroos, while shimmering grasses wave through the heat haze into the far distance.

Australia contains an enormous range of homes for its plants and animals. It is also an island, cut off from the rest of the world by the surrounding seas. It shares some of its animals with New Guinea, such as the cuscus, but most of the plant and animal species are found only in Australia.

There are many ideas about why this is so but the most likely is that Australia once shared many of its species with the rest of the world. Then, many thousands of years ago, the seas rose and cut off the land bridge between Australia and what is now known as the continent of Asia. The animals that were in Australia then adapted, over thousands of years, to the environment around them.

There are several unusual features about the animals that live in Australia. One is that many of the animals are marsupials—that is, they are mammals which raise their young in a pouch. Another is that, with the exception of the dingo, there are no large animals which feed on other animals.

Both of these unusual features exist because Australia was cut off from the rest of the world long ago. It is thought that marsupials were once found in many parts of the world but that they became extinct because they could not compete with other animals (marsupial opossums, for example, are still found in parts of the American continent).

Australia was a land with a wide range of habitats and no large predators (hunting and killing animals) and so the marsupials had plenty of places to use as homes and no real competition from other animals. The dingo had little effect because it came to Australia long after the other species had adapted to the Australian environment.

The first human settlement in Australia also had no major effect. The Aboriginal people learned to live within the Australian environment and use its plant and animal products without causing widespread destruction. It is likely that some animal species were affected by the arrival of the first human settlers but most of the animals either adapted to the change or were not affected at all.

Some people who have studied the Aboriginal way of life believe that the practice of lighting fires to clear bush or catch food may have changed some of the bush and grass areas. Other people say that the Aboriginal population was too small to have made big changes. At least it would be true to say that the Aboriginal people and the plants and animals could have continued to live together for many thousands of years without huge changes occurring.

Unfortunately this was not to be so because the European settlers arrived. Australia was a country which was totally different from the one from which they had come, so the European settlers saw many of the plants and animals as nuisances or things to be feared and destroyed.

A lack of understanding of the way the various parts of the environment worked together caused silly mistakes to be made. Farmers cleared huge areas of rainforest in the belief that the soil which grew such huge trees must be very rich indeed. It was not the soil which was rich, but the combination of plants, soil and climate.

When the rainforest was removed, the soil soon became very poor and useless for farming.

We still have a lot to learn about the Australian environment. The main thing we must understand is that each plant and animal has a part to play in not only helping the Australian environment continue to survive but also in making our lives happier and healthier.

CORAL REEFS

The pounding of a huge surf shakes the outer edge of the coral reef around the lagoon. Waves driven across the ocean by wind and current smash against the reef and spill over to die and become a swirling, white froth in the lagoon. In the shallow water along the inner edge of the reef a sea slug called a Spanish dancer swims by, the beautifully-coloured fringes of its body moving gently in the soft motion of the waves.

Gannets

In a deeper part of the lagoon a dugong glides just below the surface. A smaller shadow below shows that it is a mother with her baby. As the mother pauses to feed on sea lettuce the baby darts in to feed at the nipples under her mother's front flippers.

Amongst the brilliant-white coral sand of a nearby cay, terns and gannets squabble and fuss as they sort out their nesting territories. In one part of the island, where the sand is a little higher, pandanus palms and pisonia trees are used as nesting places by other birds. White-capped noddies, nesting in clusters amongst the sticky pisonia fruits, peer out through the foliage as frigate birds soar overhead, waiting to steal fish from gannets returning from the hunt.

Out on the reef herons strut and peck as the tide recedes. A green turtle swimming just below the surface causes a flurry of splashing as it swims through a school of small fish. Below the water tiny coral polyps duck back into the rock-like home they have made out of the sea's minerals and gradually extend their tentacles as the danger passes.

Tony Oliver 84.

An anemone fish, its brilliant flash of blue catching the soft light of the sun through the water, peers out from the tentacles of its anemone home. It sees a slow movement on a piece of brain coral as a crown-of-thorns starfish moves its tube feet over the surface of the coral feeding on the polyps as it moves. At the base of the brain coral a rock-like shape suddenly seems to have grown an eye and spines as a stonefish moves slightly to be in a better position to feed.

With the falling tide wide patches of reef are revealed, leaving deeper pools as shelter for larger animals. In one large pool a sea-snake has been caught by the falling tide and settles to the bottom to wait for the return of deeper water. Out of a deep crevice in the coral an eel stares at the snake and opens and closes its mouth to reveal needle-like teeth. The snake is neither food nor danger so the eel returns to its dark lair. In the deeper channel, schools of emperor fish sway and move in the current as they search for food.

In the shallow water where the white sand meets the coral reef there is a clutter of broken coral and empty shells. Suddenly one of the empty shells grows legs and begins to move. It is a hermit crab which has taken the empty shell of a dead shellfish as its home. Other crabs, which cannot carry their home with them, feed at the edge of burrows and cracks in the sand and reef, ready to duck below if there seems to be danger.

Death can come in many forms on a coral reef. Some beautiful cone shells have poisonous darts like small hypodermic needles with which they inject poison into their prey. The beautiful butterfly cod carries many poisonous spikes along its back fin. Sea-snakes, stonefish—death seems to be everywhere. More importantly there is warmth from the sun, huge amounts of food of all sorts, and safety and shelter for the animals who make the reef their home.

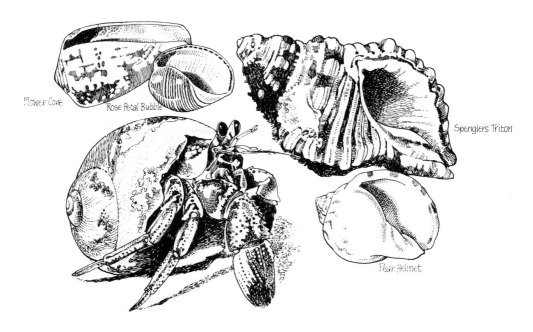

SEASHORES

Tossed high by a wave, a worn piece of driftwood clatters on to periwinkles clustered at the splash line. The force of the wave is broken by the rock shelf and the water spreads to overflow out of sparkling rock-pools. Along a crevice the waving arms of small, speckled anemones sweep the water to catch any pieces of food the wave has brought into the pool. A small, brown octopus changes colour as it moves cautiously across the bottom of the pool to pick up a dead fish which is washing with the swirl of the wave.

Along the edge of the pool, black nerite periwinkles move slowly across the patches of green algae, using their thousands of tiny teeth to scrape the algae into their mouths. A limpet, amongst hundreds, settles into the scar it has made in the rock surface to wait for an incoming tide so that it can search out a patch of algae on which to feed. It will return to its home scar when the tide recedes.

As another wave swirls over the rock shelf the cunjevoi clustered around the pool take in and pass out water through special openings. As the water passes through, food and oxygen are taken out. Under a large rock nearby, beady eyes gleam as a small rock crab watches the water swirl by. Crabs of various sizes sidle out from their rock home to pick up plant and animal pieces left by the wave.

As the surface of the water in the pool settles after the passing of the wave, Neptune's necklace of seaweed gently waves in the current. It is firmly attached to the bottom of the pool and floats in the water by the little bubbles of air held in its stem. A large piece of bull kelp waves softly backwards and forwards through the pond. It was torn from the seabed offshore during a storm and, along with its tiny animal passengers, was carried into the pool by the waves.

The feeding rock crabs scuttle back to the crevice as a wheeling silver gull passes overhead. From its view above the rocks the gull sees the body of a starfish stranded by a wave. With a screech it banks and turns to land next to the dead starfish.

Along a nearby beach other gulls hunting for scraps at the water's edge hear the screech and lift off to join the feeding gull. The flapping, screeching

Octopus

Hermit crab with Anemone

Soldier crab

13

Tony Oliver 85

gulls disturb a gannet about to plunge into the surf to take a flashing fish. The gannet returns to its lone patrol along the first line of surf and sees a line of porpoise slicing and leaping their way through the breaking waves. The school of tailor that the gannet had been following disappears as the grinning porpoises speed through the tumbling, green water.

Along the breaking edge of the surf, shells have been dumped as the waves retreat. Amongst the whitening shells of pipis and abalones, the beautifully-marked, pumpkin-shaped remains of a sea urchin are tumbled by the frothing water. A sponge skeleton floats and twists in the froth until it catches upon a heap of bubble-weed left by a previous tide. Showing like dull, blue jewels, bluebottles stranded by the tide die and rot, their power to sting and kill small sea creatures becoming weaker as they dry in the sun.

Along the edge of the dune a warm, soft breeze moves the leaves of the white banksia and makes the thicket of fine-leafed paperbarks whisper and scrape. At the back of the dune an osprey in a tall, wind-torn eucalypt looks out over the sand, surf, animals and plants working to create the living seashore.

SWAMPS & ESTUARIES

The first thing you notice about swamps is their smell. Sometimes it is just the smell of rotting plants, at other times it is a smell very much like rotten eggs. People believe that the smell of the swamps makes them useless, unhealthy places, but if there were no swamps most of the animals and plants we know from the ocean and rivers would not be able to live. The smell of swamps is caused by the rotting of dead plants and animals. As they rot away their bodies become food for living plants and animals.

Mud Skipper

The mangroves along the banks of estuaries have their roots down in thick, grey-black mud which smells like rotten eggs. The mud is so full of the smelly gas that some mangroves have to put up breathing roots to get fresh air. The purple semaphore crab and the many-coloured fiddler crabs search for pieces of food in the mud, sorting out the best pieces with their strong nippers. Mudskippers squirm and slither on the slippery surface, their bulging eyes watching out for midges or mosquitoes that might land on the mud.

Estuaries have an opening to the sea which lets the food-rich waters of the mud flats and mangrove forests mix with the wide oceans. Fish swim in the estuaries—mullet, their silver, torpedo-shaped bodies leaping out of the water at dusk, or bream, whose sharp teeth find a way into strong oyster shells.

As the tide comes in, sea-grasses lift from the mud and move gently in the currents. Long strands of brown seaweed are lifted up in the water by small floats on their stems, their roots fixed firmly to a cockleshell in the rich mud. Mud crabs as big as dinner-plates move out of their muddy burrows when the water is deep enough to cover their bodies, searching for any food which is either dead or too slow to get away from them.

Moving with slow, careful steps, spoonbills and white ibis move along in the shallow water. The curved beaks of the ibis dig down deep into the mud searching out small crabs and shellfish. The spoonbill's beak sorts out food from mud as it sweeps back and forth,

Mud Crab

dribbling out the rubbish through gaps at the end of its beak and swallowing the rest. Cormorants sit in paperbark or she-oak trees hanging their wings out to dry before returning to dive and chase and catch small fish like toadfish.

Overhead, a white-breasted sea-eagle floats and softly flaps on warm air-currents, its soft, grey and white feathers ruffling gently as it glides through the air. Its eyes search for live prey such as mullet, while high above a whistling kite gives its piercing cry as it looks for dead animals floating on the water or washed ashore.

White-faced Heron

Mangrove Goanna

In the trees along the shore a noisy argument breaks out as fruit bats fight for the best places in their roosting-trees. If the fruit bats live in the north of Australia their fights might be watched by two unblinking eyes just above the surface of the water. Salt-water crocodiles like to catch any fruit bat foolish enough to fall in the water. A distant relative of the crocodile, a mangrove goanna, searches among the buttress roots and trunks of mangroves for food and startles a white-faced heron waiting to stab a marsh crab with its beak.

Insects, reptiles, mammals, birds, fish, plants, shellfish—all the life of swamps and estuaries—depend upon the rotting plants and animals. Like the pieces of a giant jigsaw, all their lives fit together.

EUCALYPT FORESTS

It is a warm day in spring. High in the tops of the eucalypt forest a tiny pardalote sends out its double-noted call. The air is full of the smell of eucalypt oil and boronia blossoms. A lace monitor scrambles its way over rocks in sudden alarm as the shadow of a gliding goshawk shows on the ground. The monitor lies among fallen leaves and twigs, its forked tongue flicking out to find the danger. An echidna snuffles by, head down, its short legs and swinging body moving it to its favourite ant nest. It sees the movement among the leaves and suddenly digs down into the earth so that all that shows are its yellow-tipped, black spines.

Cicada

A sitella hops up the bark of a tall stringy-bark, stopping to peck out a small crusader bug hidden in the bark. A cicada clings to the bark, giving off a soft humming sound which suddenly becomes an ear-shattering thrumming. He is the first of thousands of cicadas which will appear as the weather warms. A kookaburra joins in the noise, its laughing cry telling other kookaburras that it owns the territory around the stringy-bark. It stops its cry with a gurgle and peers down at the shiny body of a skink sunning itself at the edge of its rock burrow. With a swoop and a thud of its beak the kookaburra has ended the lizard's sunbaking.

A new sound joins the others, as a raven perches on a dead branch above the kookaburra. With its neck-feathers ruffled, the raven gives a soft, gurgling cry, like that of a small baby, which changes to a harsh "caw, caw, aw, aw, aw". Other ravens join in but leave when they see that the kookaburra is not about to share its food.

As the day cools and the light dims, swamp wallabies move out of their safe, day-time homes and feed on the soft, green plants at the edge of the wet areas. A red-necked wallaby stoops to drink at a small, clear-running stream and

Bandicoot

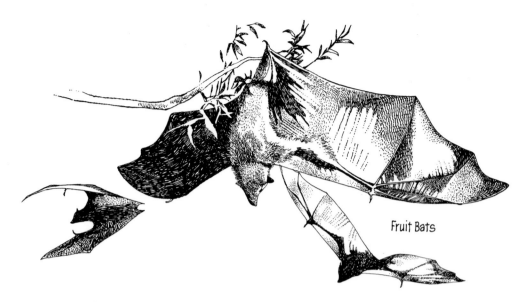

Fruit Bats

jumps back in fright as a crayfish flicks its way back to its burrow in the bank.
As the night settles, a ringtail possum peers out of its nest of bark and twigs
above the creek and then swings itself away by hand and tail to search for its
favourite leaves to eat. Two sugar gliders watch the ringtail pass, then climb
and leap and glide to a spreading banksia whose blossoms still contain some
nectar.

On the ground a bandicoot leaves its nest in kangaroo grass and moves to
the edge of the wet area where the wallabies feed. It stands still, as if listening,
and then begins to dig down with sharp claws, pushing its pointed nose into
the hole to snuffle out insects or worms.

Silently a tawny frogmouth wings its way through the darkening bush.
The sensitive hairs beside its beak seek out the movement of night-flying
moths. Deep amongst a wattle thicket the sad sound of a boobook owl carries
across the night stillness, as it calls ''mo-poke, mo-poke''. There is a clatter and
screech and a wild fluttering of leathery wings as a fruit bat lands in the top of
a flowering eucalypt. The fruit bat has found the flowering tree by following
the scent carried on the night air. It feeds hungrily on the nectar, licking and
chewing at the golden blossoms.

The movements, scents and sounds of the eucalypt forest signal to the
animals how they should act. Their lives depend upon their senses.

RAINFORESTS

The soft rustling sounds of the forest are broken by a gentle pattering, like small stones dropped on crumpled paper. It is very difficult to find where the sounds come from. Overhead the branches of the tall trees join together to form a canopy which dims the sunlight. Long vines called lianas hang down through the moist stillness and dapples of sunlight dance on the lichen- covered bark of the teak and coachwood trees.

The pattering sound comes again and small objects fall from the canopy above. It is caused by seeds and pieces of fruit being dropped by feeding animals high in the branches.

Seeing animals in rainforests is always difficult because of the lack of light and the height of the main trees. The animal feeding on rainforest figs would most probably be one of the pigeon family. Although rainforest pigeons belong to the same family as the pigeons seen in parks and gardens they are very different in behaviour and colouring. The wompoo pigeon and the purple-crowned pigeon are beautifully coloured but their colours blend with the light and shade of the rainforest. They feed quietly among the fruit-laden branches of the trees. Sometimes the stillness is broken by the echoing call of the wompoo. In times gone by hunters in the Australian rainforests found pigeons not only by sight but by their calls.

Countless animals live in and on the leaf litter where the fruit pieces fall. Some are very large animals like the amethystine python which glides silently over fallen logs and dead leaves, with its flickering tongue searching for signs of prey. Many animals are so tiny that they cannot be seen without a magnifying glass. These tiny animals—springtails, amphipods, slaters, spiders and mites—live in the leaf litter on the forest floor. They eat the decaying plant and animal matter and help break it down. The rainforest plants in turn use this decaying matter as food. Plants called fungi, and bacteria, also help to recycle the dead plants and animals. The goodness contained in their bodies can then be used again by the rainforest.

Other animals scuttle and leap about their rainforest home. Lizards in all shapes and sizes live in the parts of the forest which best suit their size, colour and choice of food. Land mullets, glistening grey-black, torpedo-shaped lizards, scuttle quickly over the forest floor on stubby legs as they search for the snails,

Amethystine Python

Great Barred River Frog

25

Tony Oliver

earthworms and other small animals which are their food. Geckos dart in and out of their bark homes. An angle-headed lizard clings to the side of a tree. The bark matches its colouring perfectly. Mammals search out food at each level of the rainforest, from under the litter to the top of the canopy. Long-nosed bandicoots dig and snuffle down into the rich litter, feeding on any small animals they can catch. A very speedy, small marsupial mouse called a brown antechinus moves around the forest floor. Like a brown blur it catches and eats spiders and insects. A tiger quoll sleepily blinks its eyes in the daylight and then returns to its hollow log to wait for night when it can hunt the small animals which are its prey. A parma wallaby lifts its head from where it is feeding on small plants at the edge of the trees and then disappears like a brown shadow into the safety of the forest.

Rainforests are mysterious places full of hidden life and movement. Every part depends on every other part to live.

Giant Snail

King Cricket

OPEN WOODLANDS & PLAINS

In the distance the tops of yellow box trees seem to swim in lakes of shining water. Halfway to the horizon, emus walking through metre-high Mitchell grass seem to be cut in half as they walk into the ''lake'' caused by the heat haze. Startled by the slowly-wheeling shape of a wedgetail eagle, the flock of emus burst out of the grass and stride over bare clay patches, making little spurts of dust as their feet strike the ground.

Under the shade of the spreading branches of a coolibah tree, a male red kangaroo lifts its head to watch the running emus. He pushes himself to his feet with his short front legs and grunts to warn the other kangaroos in his mob. Soon the whole mob is standing and unsure what danger threatens. They hop gracefully through the mulga and saltbush, heading for another favourite resting- place amongst the tall river red gums along the river bank.

The stillness of the plain is broken by a shrieking flock of budgerigars which lift in a group from a small waterhole amongst the lignum. They fly to the bare branches of a dead ironbark and perch like flickering, chattering fruit on every spare space. Around the edge of the pool they have left, zebra finches, their colours showing like small jewels, drink thirstily before flying off to feed on grass seeds.

In the waving heat haze of the late afternoon, a shingleback lizard flicks its dark-blue tongue as it moves slowly amongst the fallen leaves of a cypress pine thicket. At the edge of the sand patch where the cypress grow, a shape like a shiny, brown strap slides silently from shelter under a log. The brown snake leaves its winding trail in the soft dust as it unblinkingly watches crested pigeons dust-bathing and feeding under a wattle thicket. One bird flicks its crest and then, with a sudden whirr of wings, the small flock dips and swoops away to a safer spot.

Red-Kangaroo

Shingleback Lizard

Mallee-fowl

The mob of kangaroos hop under the shadow of the spreading river red gums. The wheezing cough of the two-metre-tall male and the thudding of the mob's hopping feet startles a flock of grey teal and black duck out of their favourite waterhole in the riverbed. In a deep part of the pool large bubbles rise to the top of the water as a long-necked tortoise moves across the muddy bottom searching for water plants. As the kangaroos settle down in the shade, galahs in a dead tree screech at a mob of cockatiels trying to share their space in the tree. With pink wings and bodies showing as they turn and dive, the galahs screech overhead to land at the water's edge to drink.

As the sun moves down in the west, the haze slowly disappears. The twittering of the zebra finches seem to get louder and a soft, buzzing sound comes from a nest of native bees hidden in a rotten cypress log. An echidna leaves its cool home under leaves and bark at the edge of a mallee thicket and snuffles its way to the heaped mound of earth which marks an ant nest. Deep in the mallee thicket a pair of mallee fowl scrape dry leaves and twigs onto a mound which is already more than one metre tall. Soon they will dig down into the hot centre of the mound and the female will lay her eggs to be incubated by the warmth of the rotting vegetation.

As the sun disappears behind the flat rim of the land and darkness descends, a strong cry like a lost spirit comes across the plain. A stone-curlew has survived another day of being hunted by fox and feral cat.

DESERTS

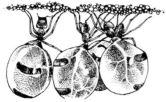

Honey-pot Ants

As the sun's heat leaves the desert rocks and night falls, small movements begin in bare patches of ground. Worker ants from a colony of honey-pot ants begin their search for nectar to store in the bodies of special workers in their nest. There is a flicker of movement as a desert skink swallows an ant and then watches as the lid of a trapdoor spider burrow slowly rises.

Two hairy legs appear on the edge of the burrow and the skink darts forward. Like the blink of an eye the lid is closed and the spider is safe underground.

Under a hard clay hollow which held water in the rare times that it rained, a water-holding frog lies curled up in a mud ball. The inside walls of the mud ball stay moist and the frog's body holds enough water for it to live until the next rain. A kowari darts across the hollow to snatch a scuttling scorpion, giving the animal a quick flick to remove its poisonous sting. A beautiful bilby emerges from its burrow under a spinifex clump and holds its long, rabbit-like ears up to catch any sound. During the day its ears help to cool its blood but at night they sense out danger from dingos or feral cats.

Water-holding Frog

As a grey, cold dawn changes into a fiery-red day, sounds, like the babbling of many voices, come from under a clump of wilga trees. Babblers, grey, busy birds the size of mynas, scratch and squabble as they hunt for food. A small patch of water, protected from the heat of the sun under a clump of bull oaks and held in a basin of rock, attracts animals from near and far.

Parrots and cockatoos screech their welcome to the day and sit around the edge of the water like many-coloured petals on a flower. A flash of crimson in the saltbush shows the flight of a crimson chat as it comes to the water. A

Nightjar

Tony Oliver

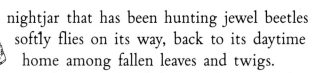
Thorny Devil

nightjar that has been hunting jewel beetles softly flies on its way, back to its daytime home among fallen leaves and twigs.

The warmth of the sun draws reptiles out of their homes in cracks in the soil and from under rocks and clumps of bushes. They have to hunt before the sun's heat drives them back to shelter. A thorny devil, looking like a very small, many-spiked dinosaur, lies in wait for the black ants which are its food. As it warms in the sun its movements become freer so that it can dart about to find new ant-trails. A sand goanna winds through clumps of blue-bush and crosses a patch of sand which is already beginning to take in the heat of the sun. It lifts its body slightly off the ground, making it look less like a thick whip and more like the long-legged, fast-moving animal it can become. A grass wren sitting on the top of a spinifex bush gives its high, peeping alarm call as the goanna moves towards the clump of spinifex where her nest is hidden. But the goanna moves away to search for larger prey, such as a marsupial mouse that might be late returning to its burrow.

The glaring light of the mid-morning sun throws a fluttering shadow on the ground. Overhead a kestrel holds its hovering position in the sky by rapidly flapping its wings. It sees a slight movement amongst a clump of canegrass. It is a hopping mouse clearing some dirt from the edge of its deep burrow before going below. The kestrel dives like a brown blur to grasp the mouse in its talons and fly to a dead branch to skin and eat its prey.

Life in the desert is one of burning heat and freezing cold. Animals and plants which survive have adapted to the harsh conditions.

Striated Grass-Wren

Greater Glider

TABLELANDS & ALPS

Each year thousands of people go to the Australian alps
to enjoy the snow. Not so long ago, Aboriginal people used
to go to parts of the alps in the warm weather to feast on
the bogong moths which gathered in great numbers at certain
times of the year.

The life of the alps is tied to the seasons, with the heaths
bursting into flower at the end of the snow thaw. The pretty-
faced wallabies move onto new, green tips as the sun-warmed
slopes lose their snow and the lumbering wombat bulldozes its
way through patches of snow as it searches for roots and
plants amongst the snowgrass and kangaroo grass. As the
weather warms, snow gums lose the last of the icy patches
around their shadowed bases and their twisted branches paint patterns along the
slopes.

Above and below the snowline and in the heaths and forest of the
tablelands, life has become adapted to harsh seasons. The matted plants of the
herbfields and the daisies of the tall herbfields cover the slopes with colour as
they bloom thickly and quickly to fit their lives into the short alpine summer.
Alpine marsh marigolds do not wait for all the snow to go before they flower
but push their pale, waxy sepals up through snow and rushing thaw water.

Bogs and fens are swampy areas which form along the banks of streams
and creeks. Corroboree frogs breed quickly in the short summer and live
protected by sphagnum moss growing along the creek beds. Glistening in
yellow and black stripes, they hide away in the moss, finding places to lay the
eggs which grow into tadpoles within the moss.

Forests of alpine ash and manna gum, with their tall, straight trunks

37

Tony Oliver 84

Chestnut Teal

reaching to the sky, make homes for gliders and possums. At night the very rare mountain pygmy possum hides and scampers in the trees and underbrush while it searches for a hollow in which to make a nest for its family. A greater glider peers out of a treetop and then leaps and glides to another tree fifty metres away. Bush rats scuttle amongst the fallen leaves and branches searching for soft plants and small animals on which to feed. Large eyes and twitching whiskers help guide them through the dark, moist undergrowth.

As the eastern sky changes to a soft grey-pink, the brown, shadowy shape of a lyrebird moves through the leaf litter. It stops to scratch at the litter, turning it over with sweeping movements of its claws and pecking at the leaping and crawling soil-animals it finds. From overhead there is a sound like a door with rusty hinges being opened slowly. A male gang-gang parrot lifts the crimson crest on its head as it lands on the dead branch of a mountain ash. As a paler female flies past, the gang-gang gives the croaking, grating cry that echoes over the forest. A wheeling flash of green and red signals the arrival of king parrots to feed on the blossoms of the manna gum.

In an icy stream, winding its gurgling way through the forest floor, young rainbow trout feed upon flying insects that fall on the water surface. A scarlet robin flits overhead, catching insects on the wing and then returning to a low branch to twitch its tail and display its flashing, red breast. In a wider part of the stream a family of chestnut teal paddle softly from the shelter of one patch of reed to the next.

Mountain animals like the pygmy possum need special homes and conditions to live. Clearing of land and carelessness when lighting fires destroys their chances to live.

URBAN

The clatter of dustbins disturbs a sleeping dog and it begins to bark. Down the street another dog answers and a chorus of howls and yips begins. A black and white tomcat, halfway along a paling fence almost slips in fright at the sudden sounds of barking. Tail flicking from side to side, its green eyes glare around to see whether it is in danger. The dogs do not even know it is there so the cat glides on to continue its search for friend or food.

City backyards at night do not only contain the noisy dog and the searching cat. Brushtail possums have learnt to live with people and their pets and they find houses very good places to use as homes. A small space under the eaves, a missing tile and a wandering brushtail sets up house above the ceiling. Hollows in logs or trees were the usual home of brushtails but city suburbs have few of those left, so the possum makes its home in the next best thing. Brushtails in the city seem to have a perfect life—finding food scraps and fruit trees at night, and sleeping through the day in a safe, warm and dry spot above the ceiling of a house. Safe, that is, until the human owner becomes tired of the noisy, uninvited guest.

While the brushtail sleeps, the urban backyard is alive with animals above and below the ground. Sparrows, starlings, and Indian mynas strut and flitter over the lawns. Snails glide back to their dark, moist homes under brickwork or in rockeries. Earthworms burrow their way through the soil. A blue-tongue lizard lies in a patch of sunlight before beginning its search for any snails that have been too slow to return to their safe homes. Overhead, blue triangle butterflies flit among the blossoms of the soft-blue jacaranda flowers, swooping to avoid the cleverly-spun web of the orb-weaving spider.

Indian Myna

41

Tony Oliver 85.

Humans are not the only animals which share the inside of the house. Huntsman spiders lie in wait in dark corners and folded curtains hoping for a fly or mosquito to land nearby. Cockroaches and silverfish hide in the dark corners of cupboards waiting for night to find scraps of food or paper. A house mouse snuggles into its nest under the floorboards waiting for dark so that it can nose out any food.

A harsh screech in the blackbutt in the front yard tells that a sulphur-crested cockatoo has arrived. Rainbow lorikeets wheel and dip, as they screech their way to an Illawarra flame-tree laden with flowers. The yellow eye of a currawong peers down at the magpie pecking a piece of meat left by the dog but the currawong decides that easier food can be found elsewhere.

Each animal has its home. Some cockroaches live only on camellia trees and some tiny insects live only on wattle flowers. Some animals, like the red-back spider, can live anywhere that provides enough shelter. Others, like the koala, need special food and shelter to survive.

Under bricks and twigs beside a paling fence, grey, many-legged slaters hide away from the daylight. They share their dark, moist homes with slugs, feeding on the rotting wood and grass of their hiding place. In cracks and crannies of the fence, common brown skinks peer out and then move to bask in the sunlight.

A prowling magpie eyes the basking lizards and pecks suddenly at the nearest. The magpie is left with a twitching, twisting tail in its beak as the sun-warm lizard sheds its tail and scuttles back to its crack in the fence.

Urban animals have to fit their way of life to the changes that people have made to the environment.

Magpie

NOCTURNAL

Moonlight is a thing of magic—soft, gentle light, often scattered by scudding clouds, adding an eerie dimension to the noctural world. For some animals night-time is a time to wake up, to hunt for food and to find a mate.

Night-time has many faces—it is the ghostly white shape of a barn owl; it is a furry, grey body moving through a eucalypt, or a slowly-flapping flying fox crossing the starlight. Barn owls and tawny frogmouths begin their hunting. The darkness hides them from their enemies and makes it easier to find food.

Noctural mammals show their adaptations in large eyes which collect any light available and in colours which let them blend into the grey world of moonlight and starlight. Large brushtail possums, small ringtail possums and delicate gliders, climb and glide through the bush looking for fruit, leaves and tree sap. On the ground, bandicoots snuffle around hunting out beetle larvae and earthworms.

It is not only the large animals which come out at night. Snails and slugs, insects, spiders, frogs—all find in the dark a safer, easier time to move about. Spiders build their wonderful webs to snare careless insects, and snails and slugs munch away with their thousands of tiny teeth.

Australia's climate is very suitable for noctural animals. For much of the year, night-time is the most pleasant time to be moving about, in contrast to the searing heat of daytime.

READING LIST

The Australian Environment series. Hodder & Stoughton Australia
CAYLEY, N.W. *What Bird is That?* Angus & Robertson
CLYNE, D. *Nature City* series. Methuen Australia
COGGER, H. *Reptiles and Amphibians of Australia* A.H. & A.W. Reed
C.S.I.R.O. *Insects of Australia* Melbourne University Press
FRITH, H.J. *Wildlife Conservation* Angus & Robertson
FRITH, H.J. & CALABY, J.H. *Kangaroos* Cheshire
GOODSIR, D. *The Gould League Book of Australian Birds* Golden Press
HADDON, F.W. *The Gould League Book of Australian Endangered Wildlife* Golden Press
McDONALD, J.D. *Birds of Australia* A.H. & A.W. Reed
McPHEE, D. *The Observers Book of Snakes & Lizards of Australia* Methuen Australia
OVINGTON, D. *Australian Endangered Species* Cassell
PIZZEY, G. *A Field Guide to the Birds of Australia* Collins
READERS DIGEST *Complete Book of Australian Birds* Readers Digest
RIDE, W.D.L. *A Guide to the Native Animals of Australia* Oxford University Press
ROLLS, E. *They All Ran Wild* Angus & Robertson
SLATER, P. *A Field Guide to Australian Birds* Rigby
STRAHAN, R. ed. *The Complete Book of Australian Mammals* Angus & Robertson
TROUGHTON, E. *Furred Animals of Australia* Angus & Robertson

INDEX